Influence of Prosopis Juliflora Wood Flour in Poly Lactic Acid – Developing a Novel Bio-Wood Plastic Composite

Sachin Sumathy Raj
Thanneerpanthalpalayam Kandasamy Kannan
Rathanasamy Rajasekar

ELIVA PRESS

ELIVA PRESS

Sachin Sumathy Raj

Thanneerpanthalpalayam Kandasamy Kannan

Rathanasamy Rajasekar

A Bio composite comprising Prosopis Juliflora Fiber (PJF) and Poly Lactic Acid (PLA) was processed considering two particulate sized reinforcements, coarse PJF (avg. 15 μm) and fine PJF (10-50 nm). They were added individually at ratios of 10, 15, 20 and 25 wt% into PLA matrix. The composites were extruded and tested for mechanical properties. The addition of PJF resulted with an increase in the tensile, flexural and impact strengths of the polymer. Adding PJF to PLA showed a decrease in the hardness of the polymer. Water Absorption test showed an increase in water uptake with increasing fiber content. The most optimum ratio of PLA to PJF was found to be 80:20. The fine PJF reinforced composites proved to be superior over the coarse PJF reinforced composites at all stages of the research. FESEM and TGA were used to study morphology and thermal characteristics respectively.

Published by Eliva Press SRL
Address: MD-2060, bd.Cuza-Voda, 1/4, of. 21 Chişinău, Republica Moldova
Email: info@elivapress.com
Website: www.elivapress.com

ISBN: 978-1-63648-119-7

INFLUENCE OF PROSOPIS JULIFLORA WOOD FLOUR IN POLY LACTIC ACID– DEVELOPING A NOVEL BIO-WOOD PLASTIC COMPOSITE

Sachin S Raj[1]*, T K Kannan[2] and R Rajasekar[3]

[1,2]*Department of Mechanical Engineering, Gnanamani College of Technology, 637018 Namakkal, Tamilnadu, India)*

[3] *Kongu Engineering College (Department of Mechanical Engineering, 638060 Erode, Tamilnadu, India)*

*email : sachinsraj1991@gmail.com

Abstract

A Bio composite comprising Prosopis Juliflora Fiber (PJF) and Poly Lactic Acid (PLA) was processed considering two particulate sized reinforcements, coarse PJF (avg. 15 μm) and fine PJF (10-50 nm). They were added individually at ratios of 10, 15, 20 and 25 wt% into PLA matrix. The composites were extruded and tested for mechanical properties. The addition of PJF resulted with an increase in the tensile, flexural and impact strengths of the polymer. Adding PJF to PLA showed a decrease in the hardness of the polymer. Water Absorption test showed an increase in water uptake with increasing fiber content. The most optimum ratio of PLA to PJF was found to be 80:20. The fine PJF reinforced composites proved to be superior over the coarse PJF reinforced composites at all stages of the research. FESEM and TGA were used to study morphology and thermal characteristics respectively.

Keywords: Biocomposite material, Poly Lactic Acid, Prosopis Juliflora, Wood flour, Wood Plastic Composite.

TABLE OF CONTENTS

Chapter no.		Content	Page. No.
		Abstract	1
1		**Introduction**	3
2		**Materials and Methods**	4
	2.1	Matrix Material	4
	2.2	Fiber Extraction	5
	2.3	Preparation of PJF reinforced PLA composites	6
	2.4	Mechanical Characterization	9
	2.5	Morphological Analysis	10
	2.6	Degradation Studies	10
3		**Results and Discussions**	10
	3.1	Tensile Properties	10
	3.2	Flexural Properties	13
	3.3	Impact Properties	14
	3.4	Hardness	16
	3.5	Water Absoprtion Test	17
	3.6	Morphological Studies	22
	3.7	Thermogravimetric Annalysis	24
4		**Conclusions**	25
		References	26

1. Introduction

Non biodegradable polymers and composites have always been a difficult task when it comes to waste management, they are not decomposable and pose a major threat towards land pollution. Biodegradable thermo plastics on the other hand, plastic materials derived commonly from agro products like cassava, sugarcane and beet[1]. These bioplastics undergo complete decomposition when they are buried, thereby helping to avoid land pollution. The rising demand for biomaterials as an alternative to non-biodegradable plastics has led to vast research to develop biopolymer composite materials that possess excellent mechanical properties and are suitable for various applications.

This research involves on such investigation, where Prosopis Juliflora Fiber (PJF) is reinforced in coarse and fine particle forms individually into Poly Lactic Acid (PLA) matrix to develop a novel bio-composite material for Wood Plastic Composite (WPC) applications, focusing on good flexural and impact properties along with low water absorptiona and good thermal stability.

PLA, a starch based biodegradable thermoplastic polymer is one of the most widely used bio-plastics in the world. Experiments with wood flour reinforced into different biopolymers[2,3] have been successfully executed among which PLA based composites have shown outstanding mechanical performance. PLA has also proven to have better mechanical properties over conventionally used petroleum based polymers like Polypropylene[4] & Polystyrene[5] and can also be processed by similar methods. PLA and its composites are currently being used in versatile applications like aircraft and automobile interiors, medical implants, 3D printing and other biomedical equipment[4,5].

Prosopis Juliflora (PJ) a medium sized tree found in the tropical zones around the globe and abundantly available in South Asia, South America and Africa. It is considered as a weed to be eradicated due to its abnormal water absorbing tendency[6]. PJ is currently used as fire wood and its fruit as animal fodder. Saravanan et al.[7] did a detailed study on PJF, and has proven its superior chemical properties when compared to other commonly used plant fiber reinforcements like jute, flax, ramie, hemp, kenaf, and okra. PJF has one of the highest Lignin contents among plant fibers and this chemical acts as a thermal stability compound. Lignin also helps in providing higher stiffness and as a water proofing agent in the micro fibrils[8,9,10]. Plants with higher lignin content also have a natural capability of resistance towards brown and white rot fungus, which are common wood fungal attacks[8]. These features of PJF led to this approach of reinforcing this economically cheap source of wood fiber into PLA matrix to develop the novel PLA/PJF Bio-composite material, a combination that has not been taken up by any research till date to the author's best knowledge. PJF also having one of the lowest wood fiber densities[7] ensures to provide a light weighing WPC. Literature reveals that PLA has been previously worked with a few saw dust reinforcements like rubber wood[11], Poplar wood[12,13] and maple wood[14], which have showed improved mechanical properties. In few other cases, wood filler particles like bamboo[15] and pine[16] had shown reduction in the mechanical properties of PLA when reinforced. PJF when reinforced into Epoxy had produced improved mechanical properties of the polymer[17].

2. Materials and Methods

2.1. Matrix material

PLA of grade 3052D is one of the most widely used bio-thermoplastic polymer and was obtained from Natur tek, Chennai, Tamilnadu, India. PLA was obtained in the form of granules having a density of 1.24 $g.m^{-3}$, a glass transition temperature of 55-60 °C, a crystalline melt temperature of 150-160 °C and a melting temperature ranging between 170 and 180 °C. The granules were bright white in color in their virgin form.

PJ wood was obtained from the trees at the waste lands of Namakkal district, Tamilnadu, India. The outer layers were peeled off and the inner solid bark was dried in an oven at 90 °C for 12 hours. The bark was then held in the chuck of a lathe machine and turning operation was performed (500 rpm and 2mm depth of cut) to obtain long continuous fiber strings as shown in Figure 1a.

(a) (b)

(c) (d)

Figure 1.   a) Turning operation carried out on the lathe machine to extract the fibers from the bark, b) Alkali treated and dried fibers in the form of long strands, c) Coarse fiber particles (magnification 10 X), d) Fine fiber particles (magnification 10 X).

The wood fiber strings were alkali treated for 12 hours in a solution containing 95% water and 5 % NaOH[18, 19] to enhance the adhesion of the fiber with the matrix. The soaked fiber was again dried in an oven at 90 °C for 12 hours to remove all the moisture as shown in Figure 1b. Finally the dried fibers were powdered in a pulverizer (an equipment commonly used to crush large wood pieces into powder). The powdered PJF was sieved using a #400 mesh to obtain the Coarse PJ reinforcement as shown in Figure 1c. The average size (a particle between the smallest and largest size was considered) of the coarse powder was around 12 μm which was measured using FESEM at 4000 X magnification, (Model: Zeiss, Coimbatore Institute of Technology, Coimbatore) as shown in Figure 2a. A portion of the coarse powder was further ground into fine particles using a food processer at 10000 rpm approximately. After every two minutes of grinding, the lid of the processor was opened and the wood flour that had been accumulated on the lid's inner surface due to centrifugal force, was collected. This procedure was followed to replicate the study carried out by Fan[21], where the author had used the principle of centrifugal force and additional setups in a pulverizer to achieve superfine wood particles. These fine particles were considered as the nano PJ reinforcement as shown in Figure 1d, having a size ranging from 10 to 50 nm (measured using Particle Size Analyzer, Model: Nanophox, Nano Lab, KSR Engg College, Erode, India). The graphical output of the nano particle size analysis is shown in Figure 2b.

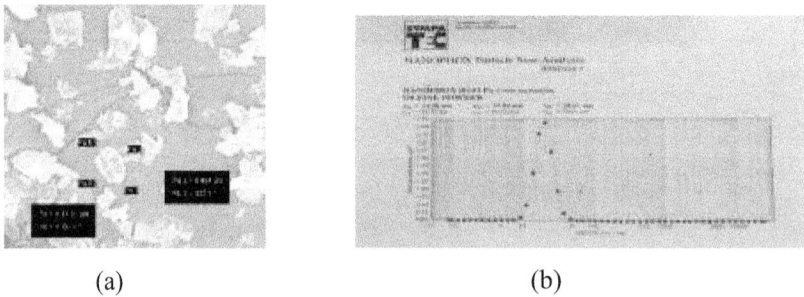

(a)   (b)

Figure 2.   a) Size of the coarse particles, measured using FESEM and b) Size of the fine   particles, measured using Nano particle size analyzer.

2.3. Preparation of PJF reinforced PLA composites

Measured quantity of PLA and wood flour as shown in Table 1 were processed using a mini twin screw extruder (Kongu Engineering College, R&D,

Tamilnadu, India) which had four processing stages. PLA was fed into stage 1 (170 °C), through a hopper, wood flour was added at stage 2 (180 °C) through a side feeder. Stage 3 (190 °C) ensured thorough belnding of the matrix and filler. Stage 4 (170 °C) near the exit nozzle ensured that the molten composite was extruded through dies of ASTM dimensions. The extruded composite bars were then cut into 10 mm thick test specimens for tensile test (ASTM D256) and flexural test (ASTM D790). The impact test (ASTM D256) specimens were obtained from cutting the length of additional fleural specimens to 127 mm.

Table 1:   Composition of PLA to PJF in wt% ratio for coarse and fine fiber reinforced composites.

Fiber form - collective name	Name of the composite	wt% of PLA matrix	wt% of fiber reinforced
Plain Polymer	PLA	100	0
Coarse/Micro sized PJF reinforced samples. – "C Samples"	C1	90	10
	C2	85	15
	C3	80	20
	C4	75	25
Fine/Nano sized PJF reinforced samples. – "F Samples"	F1	90	10
	F2	85	15
	F3	80	20
	F4	75	25

Huda et al.[14] has proved that the fiber reinforcement in a polymer matrix less than 10 wt% ratio reduces the tensile strength of the composite material due to insufficient fiber loading thereby resulting in flaws or plasticization effect of the composite. Considering the former statement, the amount of reinforcement for this research was set to start from 10 wt% with an increase of 5 wt% for each consecutive composite specimen that was fabricated. Fabrication of the composite was not successful beyond 25 wt%. At 30 wt% filler content (trial attempted), the physical quantity of the filler material was greater than that of the matrix material. Extrusion, therefore could not be carried out due to the high melt viscocity. This occurrence coincides with the study carried out by Valentina[22] who had investigated the rheology of spruce flour reinforced PLA composite and proved that the melt flow index was very large at 30 wt%, thereby preventing the easy processing of the composite. This phenomenon is also justified by the fact that the density of PJ is very low when compared to other commonly used fibers like jute, ramie, flax, hemp and kenaf which is tabulated in detail by Saravanan et al.[7] who determined the density of PJF to be 580 $kg.m^{-3}$ while that of Jute, Flax, Ramie, Hemp and Kenaf fibers were 1460, 1500, 1500, 1480 and 1400 $kg.m^{-3}$ respectively. Due to the low density of PJF, the physical quantity of fiber during reinforcement was very high, thereby restricting the fabrication of the PLA/PJF composite to a maximum fiber loading limit of 25 wt%.

Chandramohan et al.[23], explained the detailed shape and dimensioning of polymer composite materials with respect to the ASTM standards that have been considered for this study. The composite materials were fabricated to ASTM D638 (Type I) for tensile test, depicted in Figure 3a, 3c and 3d. ASTM D790 was followed for flexural test specimens (Figure 3b, 3e) and ASTM D256 for impact test. These ASTM standards were also followed by various other researchers[14,16,18,24] to determine the mechanical characterization of polymer composite materials.

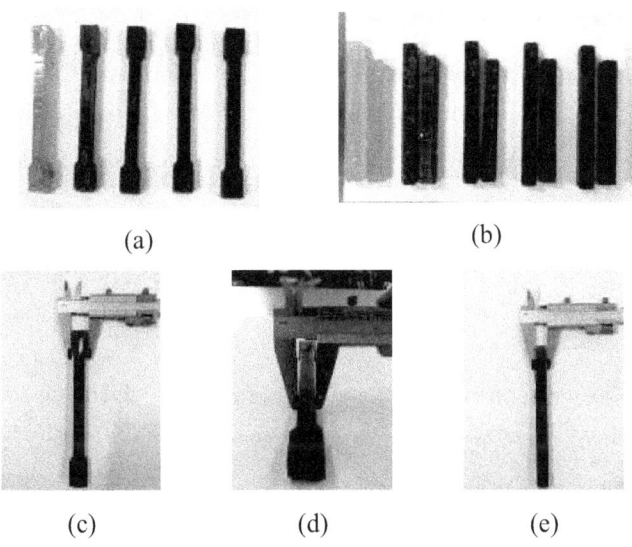

(a)   (b)

(c)   (d)   (e)

Figure 3.   a) PLA and coarse fiber reinforced composites as per ASTM D638, b) PLA and composites fabricated to ASTM D790 and ASTM D256 for flexural and impact test respectively, (c) Width of tensile specimen at the end, (d) width of tensile specimen at the neck and (e) width of the flexural specimen.

2.4 Mechanical Characterization

Mechanical testing was performed on all the speimens shown in table 1. Tensile test was carried out using a Universal Testing Machine (Brand: Kalpauk, Model: KIC-2-1000-C) operating with a load cell of 10 KN and a cross head speed of 5 mm.min^{-1}. Flexural test was also carried out on the same UTM, with the test specimen placed on a three point bending fixture, with a span of 120 mm. Izod impact test was carried out on unnotched specimens using a 15 kg hammer head weight. Vickers micro hardness test was conducted (Make: Wilson hardness, Model: 402 MVD) and Vickers Hardness Number (HV) was calculated for each of the test specimens. A load of 100 kgf was maintained as constant for all the samples, with a dwell time of 10 seconds.

The fractured surface of all the WPCs were scanned under FESEM. SEM micrographs of the flexural specimens were scanned at magnifications of 2000 X and 1000 X to obtain clear images. The scale was kept constant at 20 μm. The specimens were gold sputtered prior to FESEM.

Focusing on constructional applications for which this composite is being developed, the composites were subjected to tests relating to environmental conditions like water and heat/thermal surroundings. Water Absorption Test (WAT) was carried out since a natural fiber is used as a filler, which has a natural tendency to absorb water. Thermo Gravimetric Analysis (TGA) was conducted to analyze the thermal degradation and thermal stability. Weight measurement for the WAT was carried out using a high sensitive digital weighing scale. The specimens were first measured in their dry condition (dry weight). Then they were immersed in water at room temperature for 48 hours. The specimens were finally taken out, pat dry and measured again for the final weight (wet weight). Water Absorption was calculated using the formula[25] [(Wet weight – Dry weight) / Dry weight] x 100.

TGA was used to study the thermal stability of PLA, PJF and the WPC that resulted in providing the best mechanical property. 50 mg of samples were place in the test pan in a nitrogen atmosphere and treated upto a maximum of 500 °C with variation of 20 °C/min.

3. Results and Discussions

3.1 Tensile Properties

The tensile strength in both the coarse and fine particle reinforcements increased with increasing fiber content upto 20 wt% which had a maximum tensile strength of 21.14 MPa for C3 and 24.69 MPa for F3 while plain PLA had a tensile strength of 10.05 MPa as shown in Figure 4. PLA had an increase in tensile strength by 110% and 145% with addition of 20 wt% coarse and fine filler cases respectively. At 25 wt% loading, the tensile strength of C4 had a negligible reduction to 20.78 Mpa and F4 had a reduction to 21.66 Mpa. This may be due to the insufficient wetting of the fiber by the matrix material due to the higher physical presence of the filler material at the higher reinforcement level which was clearly evident in the FESEM image 10d. The tensile strength of F4 also had reduced when compared to the tensile strength of F3 due to the same reason. This study revealed that the F composites had superior tensile strength over the C composites in all the four cases of fiber loadings since the nano particles had better reinforcement capability than the micro particles[20]. The values of the tensile modulus followed similar fashion of the tensile strength values as shown in Figure 5a. The tensile modulus of both the C and F composites increased upto 20 wt% filler content and then reduced at 25 wt% reinfrocement. The good bonding between the matrix and fiber due to the addition of PJ may have increased the stiffness.

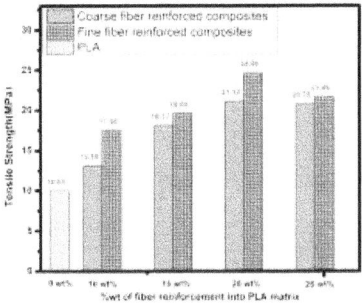

Figure 4.   Tensile strength for PLA, Coarse fiber reinforced composites and Fine fiber   reinforced composites.

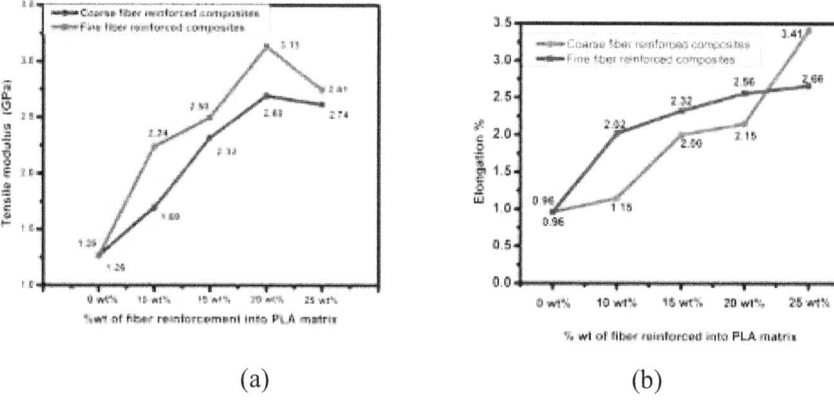

(a)   (b)

Figure 5. (a) Tensile modulus of coarse and fine fiber reinforced composites and (b) Percentage Elongation at break of coarse and fine fiber reinforced composites.

Elongation of PLA increased with increasing PJ content in both the coarse and fine cases. This may be due to the nature of PJF, which when reinforced into PLA reduced its brittleness by giving it additional plasticity. C4 had the highest elongation among all the composites. The coarse PJ reinforced composites individually showed larger variation in elongation with each consecutive sample (C1, C2, C3 and C4), while the fine PJ reinforced composites showed smaller increase in elongation values with each consecutive sample (F1, F2, F3 and F4), as shown in Figure 5b. Nasrin et al[26], carried out a study where chitin was added to PLA. An addition of chitin content from 1, 5 upto 10% into PLA matrix, showed an increase in the tensile strength, tensile modulus and elongation parallely. A study with epoxy-bagasse also showed increase in the elongation of the composite with increment in the filler content from 15 to 30 wt%[27].

3.2 Flexural Strength

The flexural strength of all the C and F composites increased with upto 20 wt% filler content, where a flexural strength of 59.95 MPa and 67.73 MPa were observed for C3 and F3 respectively. PLA had a flexural strength of 20.61 MPa as shown in Figure 6.

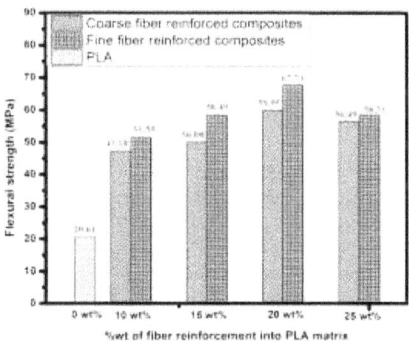

Figure 6.   Flexural strength for PLA, Coarse fiber reinforced composites and Fine fiber reinforced composites.

The maximum increase in flexural strength was 190% and 228% with addition of 20 wt% coarse and fine PJ particles, respectively, into the PLA matrix. At 25 wt% loading, the flexural strength dropped down to 56.49 MPa for the C4 sample and 58.51 MPa for the F4 sample. This was due to the dominating physical quantity of wood particle content which had eventually led to the loose bonding between the matrix and fiber. The matrix therefore could not achieve complete wetting of the fiber which inturn could not producce good fexural strength as of the 20 wt% filled composite. This study revealed that the F composites had superior flexural strength over the C composites in all the four cases of fiber loadings since smaller sized reinforcements provide better mechanical properties over larger sized reinforcement by providing larger surface area and better impergnation with the matrix material[20]. The flexural modulus increased with increasing filler content as shown in Table 2.

Table 2: Flexural modulus for PLA, Coarse fiber reinforced composites and Fine fiber reinforced composites.

Name of the composite	% of fiber reinforcement	Flexural Modulus (GPa)	Name of the composite	% of fiber reinforcement	Flexural Modulus (GPa)
PLA	0	17.70	PLA	0	17.70
C1	10	40.51	F1	10	44.26
C2	15	48.16	F2	15	50.22
C3	20	51.47	F3	20	58.16
C4	25	48.39	F4	25	50.24

3.3 Impact Strength

Impact strength of the composites were greater than that of the polymer (0.45 $J.mm^{-2}$) in all reinforcement levels as shown in Figure 7. C1, C2, C3 and C4 had an impact strength of 0.55, 0.65, 0.89 and 0.67 $J.mm^{-2}$ respectively while C4 sample had a decrease in impact strength which is justified by the SEM image studies from the delamination formed, as shown in Figure 10d. These delamination were responsible for the resistance towards the impact force. The number/area of delamination was lesser for C4 when compared to the C3 composite as show in Figure 10c. This may also be due to the dominating fiber content which was similar to the case of the reduction in the tensile strength of the C4 sample.

In the case of fine fiber reinforced composite materials, F1, F2, F3 and F4 had impact strengths of 0.60, 0.74, 1.09 and 0.92 $J.mm^{-2}$ respectively. The nano particulate fibers at the highest fiber loading of 25 wt% dominated over the plasticization effect of the matrix material to help in absorbing and transferring the energy created by the impact test to the polymer matrix effectively. This was evident through the morphological studies by comparing figure 10 e, f, g and h, which clearly showed the areas of resistance towards the shear force during the impact testing.

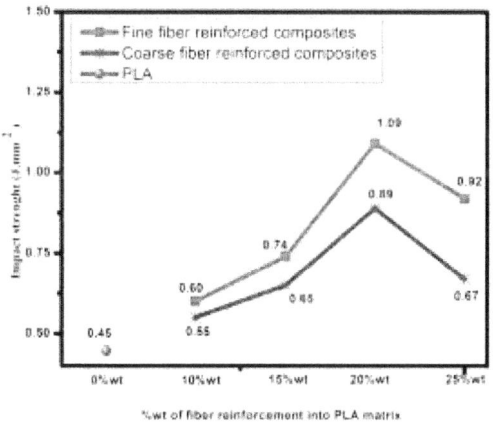

Figure 7.   Impact strength for PLA, Coarse fiber reinforced composites and Fine fiber reinforced composites.

Comparing the coarse and the fine fiber specimens, the Fine particle reinforced samples had better impact strength over the coarse reinforced samples at all the four different fiber loadings. This was due to the fact that nano particles have better distribution in a polymer matrix when compared to micro particles, which might have helped in the even energy transfer within the composite during the impact test[28]. The addition of nano particle improving the impact strength of a polymer composite was also proved by Nagalingam et al[29]. The impact resistance was found to be highest at 20 wt% fiber loading for both variants of composites.

Figure 8 shows that the Vickers Micro hardness of all the composite samples was lower than PLA in both the micro and nano reinforcement studies due to the addition of the PJ. Increase in fiber content had a reduced pattern on hardness values, the C1 and F1 composite had the highest hardness while C4 and F4 composite had the lowest hardness value within their categories. This factor is in contradiction with the fact that hardness increses with increase in elastic modulus[30]. The observed reduction in hardness can only be assumed that the nature of PJ, since every wood fiber has its own distinctive property[31]. PJF might have had the ability to enhance the rigidity of PLA by providing increase in modulus by possesing good bonding ability. On the other hand PJF had given a plasticizer effect on PLA (the latter being a very brittle material by nature[32]), thereby turning PLA into a softer compound thereby justifying the increase in the elongation as well as reduction in the hardness with increasing particle reinforcement. The hardness of the F samples was greater than the C samples at all reinforcement levels since smaller particles provide better mechanical properties when compared to the larger particles[33]. This factor was also substantiated in a study carried our by Sifat et al[34], where nano paricle reinforcement had shown better mechanical performance than the micro particles.

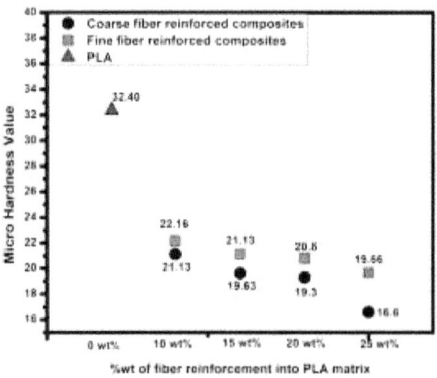

Figure 8.   Hardness for PLA, Coarse fiber reinforced composites and Fine fiber reinforced composites.

3.5 Water Absorption Analysis

The PLA sample did not show any water uptake while in both the coarse and the fine reinforced samples, an increase in fiber content showed increase in water absorption. Comparing the coarse and the fine fiber specimens, sample F1 having the least water absorption tendency among the F samples also proved superior to C1 which had the least water absorption tendency among the C samples. The C samples at all reinforcement levels had greater amount of water uptake when compared to the F samples as tabulated in Table 3. The Standard Deviation for all the mechanical properties and water absorption test that were analyzed are elaborated in Table 4.

Table 3 :   Water Absorption test results for PLA, Coarse fiber reinforced composites and Fine fiber reinforced composites.

SAMPLE	INITIAL WEIGHT OF THE SAMPLE(grams)	FINAL WEIGHT OF THE SAMPLE(grams)	% of water absorbed
PLA	3.488	3.488	0
C1	2.720	2.722	0.073
C2	2.696	2.699	0.111
C3	2.650	2.657	0.263
C4	2.629	2.638	0.341
F1	3.146	3.147	0.031
F2	3.115	3.117	0.062
F3	3.057	3.061	0.119

| **F4** | 3.024 | 3.031 | 0.198 |

Generally NaOH treatment on natural fibers lead to a phenomenon known as Super-swelling[35] which is responsible for greater water absorption tendancy of the fiber. In this research the micro PJ particles may have undergone larger super swelling than the nano particulates thereby the showing a larger physical presence of wood particles in the matrix at the similar reinforcement ratios. For better understanding a diagramatic explanation is shown in Figure 9 where PLA matrix is filled with 25 micro particles as C composites and 25 nano particles as F composites. The amount of PLA is greater in the F composites thereby leading to lower water uptake.

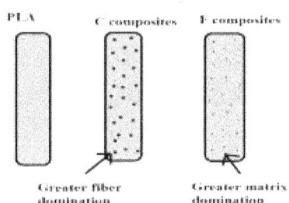

Figure 9. Diagrammatic explanation of lower micro and nano fiber distribution into PLA

Table 4:   Standard Deviations.

Property	Sample	PLA	C1	C2	C3	C4	F1	F2	F3	F4
Tensile Strength	1	9.67	13.58	18.96	20.67	21.15	17.61	19.29	24.6	20.93
	2	9.35	13.38	17.62	20.78	21.31	17.82	20.27	24.91	21.18
	3	11.02	12.98	18.53	21.56	20.36	16.98	19.89	25.52	22.21

	4	9.76	12.54	17.8	21.22	20.88	18.06	19.12	24.18	22.38
	5	10.45	13.46	17.95	21.48	20.2	17.33	19.64	24.25	21.61
	MEAN	10.05	13.19	18.17	21.14	20.78	17.56	19.64	24.69	21.66
	SD	0.67	0.43	0.56	0.4	0.48	0.42	0.46	0.49	0.56
Tensile Modulus	1	0.91	2.11	2.17	2.56	2.84	1.94	2.31	3.22	2.46
	2	1.17	1.35	2.35	2.68	2.57	2.31	2.49	3.18	2.83
	3	1.46	1.81	2.48	2.44	2.68	2.11	2.58	3.27	2.78
	4	1.53	1.61	2.1	2.96	2.63	2.45	2.38	2.88	2.46
	5	1.23	1.56	2.51	2.81	2.97	2.4	2.73	3.1	2.52
	MEAN	1.26	1.69	2.32	2.69	2.74	2.24	2.5	3.13	2.61
	SD	0.25	0.28	0.18	0.2	0.16	0.21	0.17	0.15	0.18
Elongation	1	0.91	1.18	2.01	2.11	3.36	2.1	2.42	2.49	2.63
	2	1.05	1.23	2	2.14	3.49	2.13	2.31	2.54	2.63
	3	1.01	1.11	1.91	2.17	3.42	1.95	2.29	2.55	2.65
	4	0.91	1.09	2.04	2.2	3.38	1.89	2.25	2.59	2.72
	5	0.93	1.14	2.09	2.13	3.4	2.04	2.34	2.61	2.67
	MEAN	0.96	1.15	2	2.15	3.41	2.02	2.32	2.56	2.66
	SD	0.057	0.05	0.06	0.03	0.04	0.09	0.05	0.04	0.03
Flexural Strength	1	20.08	47.12	55.4	60.21	56.68	50.87	57.85	66.96	58.54
	2	19.5	47.52	56	60.54	56.82	51.46	58.72	67.93	59.34

	3	21.87	46.86	56.79	59.66	55.67	51.3	58.19	67.71	58.9
	4	20.92	46.5	56.41	59.23	56.31	51.91	59.23	68.68	57.79
	5	20.68	47.91	55.81	60.08	57.07	52.21	58.46	67.39	57.98
	MEAN	20.61	47.18	56.08	59.95	56.49	51.55	58.49	67.73	58.51
	SD	0.89	0.55	0.53	0.51	0.54	0.52	0.52	0.57	0.57
Flexural Modulus	1	17.28	39.98	47.24	50.6	48.15	44.4	51.11	57.05	50.82
	2	16.61	40.34	47.79	51.47	47.46	43.02	50.88	57.65	49.78
	3	17.85	39.75	48.5	50.85	48.82	43.82	49.24	58.8	50.35
	4	18.32	40.86	48.22	52.45	49.11	44.85	49.57	59.1	51.05
	5	18.45	41.6	49.05	51.98	48.41	45.21	50.3	58.2	49.2
	MEAN	17.7	40.51	48.16	51.47	48.39	44.26	50.22	58.16	50.24
	SD	0.68	0.66	0.61	0.68	0.57	0.77	0.72	0.74	0.68
Impact Strength	1	0.49	0.5	0.75	0.79	0.59	0.69	0.71	1.02	0.81
	2	0.45	0.53	0.69	0.84	0.62	0.61	0.68	1.08	1.01
	3	0.38	0.6	0.61	0.94	0.69	0.51	0.75	0.98	0.97
	4	0.42	0.57	0.61	0.99	0.79	0.57	0.77	1.21	0.88
	5	0.51	0.56	0.58	0.88	0.66	0.62	0.8	1.16	0.93
	MEAN	0.45	0.55	0.65	0.89	0.67	0.6	0.74	1.09	0.92

	N									
	SD	**0.05**	**0.03**	**0.06**	**0.07**	**0.07**	**0.06**	**0.04**	**0.08**	**0.07**
Micro Hardness	1	33.5	20.95	18.9	19.18	16.41	22.66	21.28	20.65	20.02
	2	31.32	21.36	19.44	19.25	16.66	21.91	21.09	21.15	19.15
	3	31.65	21.61	19.76	18.95	16.11	22.4	21.47	21.03	19.32
	4	33.03	20.6	20.1	19.48	17	21.71	20.86	20.03	19.68
	5	32.5	21.15	19.95	19.64	16.82	22.15	20.97	20.87	20.13
	MEAN	**32.4**	**21.13**	**19.63**	**19.3**	**16.60**	**22.16**	**21.13**	**20.8**	**19.66**
	SD	0.8	0.3	0.4	0.2	0.3	0.3	0.2	0.3	0.4
Water Absorption	1	0	0.073	0.111	0.247	0.365	0..031	0.057	0.112	0.19
	2	0	0.073	0.103	0.284	0.32	0.031	0.06	0.137	0.185
	3	0	0.073	0.119	0.258	0.335	0.031	0.069	0.108	0.219
	MEAN	**0**	**0.073**	**0.111**	**0.263**	**0.34**	**0.031**	**0.062**	**0.119**	**0.198**
	SD	**0**	**0**	**0.008**	**0.019**	**0.022**	**0**	**0.006**	**0.015**	**0.018**

Figure 10 a, b , c and d show the FESEM images of C1, C2, C3 and C4. The morphological image of C1 shows large areas with insufficient PJ loading which was the reason behind the low mechanical properties exhibited by C1 when compared to the higher reinforcement content composites. C2 had better fiber distribution when compared with C1 which improved the mechanical properties over C1. C3 being the best sample among the C samples, shows clear evidence of superior mechanical properties form the large delaminates formed on its fractured surface during the testing. C4 has largely dominating fiber content and reduced amount of laminates; hence the mechanical strengths were comparatively reduced when compared to C3. Figure 10 e, f, g and h show the FESEM images of F1, F2, F3 and F4. The morphological images of the fine fiber reinforced composites had large areas displaying clear resistance to the shear force during testing. F3 sample having the highest mechanical properties when compared to F1, F2 and F4 clearly exhibits the largest area of resistance to the flexural force.

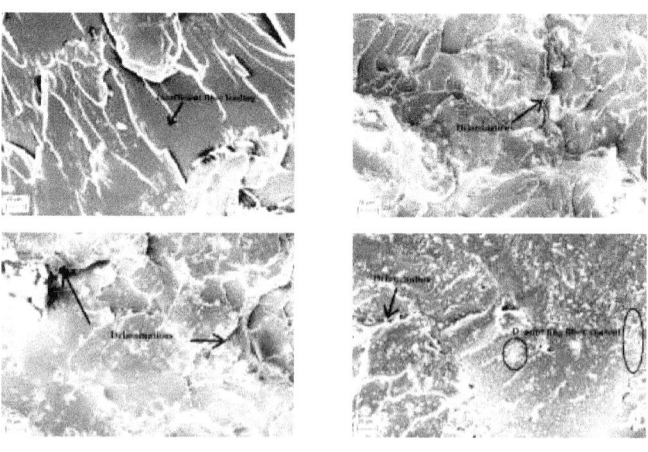

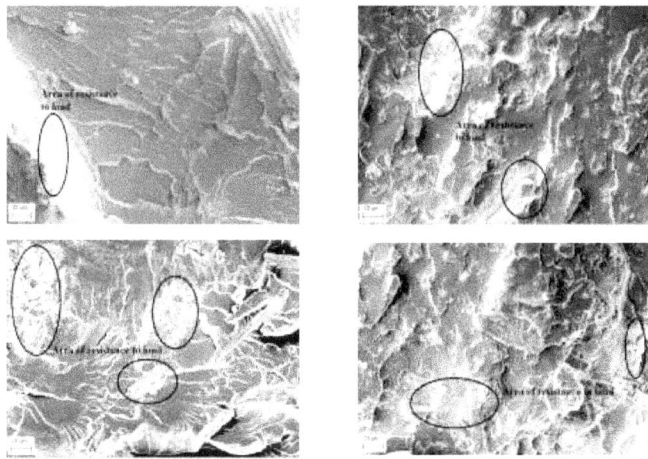

Figure 10: (a) FESEM image of C1, (b) FESEM image of C2, (c) FESEM image of C3, (d) FESEM image of C4, (e) FESEM image of F1, (f) FESEM image of F2, (g) FESEM image of F3, (h) FESEM image of F4.

The FESEM images of the 10 and 15 wt% PJ reinforced composites in both the coarse and fine cases show better delamination when compared with plain PLA. This evidently proves that the addition of PJF incresed the mechanical properties of the polymer. The increase in the mechanical properties with increasing PJF reinforcement into PLA may be due to the excellent bonding of the PJF with PLA matrix, thereby resulting in a good resistance towards the testing forces. It is also visible that PJF has even particulate distribution within the matrix material with each consecutive fiber increament upto 20 wt% beyond which agglomeration of reinforced particlutes reduces the mechanical properties. Figure 11 shows the FESEM image of plain PLA sample that had been scanned on its fractured surface after the flexural test. The very small amount of delammination in it shows that it had lower mechanical strength when compared witht the micro annd nano PJF reinforced composites.

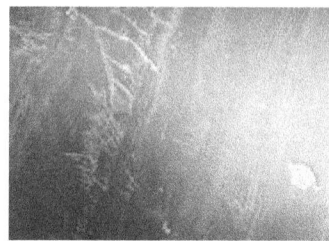

Figure 11: FESEM image of PLA

The composites in both the C and F cases with 20 wt% fiber loading had the best mechanical properties which is extensively in synchronization with a couple of researches who had used various wood fibers with PLA[12,36]. Ediga et al.[17] had analyzed the mechanical properties of Epoxy-PJF composite and also concluded that 20 wt% of PJF reinforced composite had the best strengths which strongly supports the fiber loading parameter of this study. The mechanical properties of the PJF/Epoxy composite[17] and PJF/PLA composite (this study), when compared showed that PJF performed as a better reinforcement with PLA than with epoxy.

3.7 Thermo Gravimetric Analysis (TGA)

Thermal stability of PLA was good upto 300 °C beyond which it had a sharp degradation rate and the polymer reduced to a mass of just 6% at 400 °C as shown in Figure 12. There were no residue remains at 500 °C, since PLA had turned volatile. PJF began to degrade around 200 °C losing 7% mass and was thermally stable to nearing 270 °C. Between 300 to 400 °C there was a sharp decrease in the mass to 9%. The F3 specimen had equivalent thermal stability to that of the PJF and PLA, following the rule of mixtures. Majority of the plant fibers are thermally stable upto 200 °C, while PJF had superior thermal degradation comparing to other commonly used natural fiber reinforcements. This may be due to the high lignin content in PJF which is known to be responsible for the thermal stability of natural fibers[8]. The lignin content in PJF is 17%[7] while other common plant fiber reinforcements like Flax, kenaf, jute, hemp and Sisal have a lignin content of 2, 9, 12, 10 and 9 respectively[37].

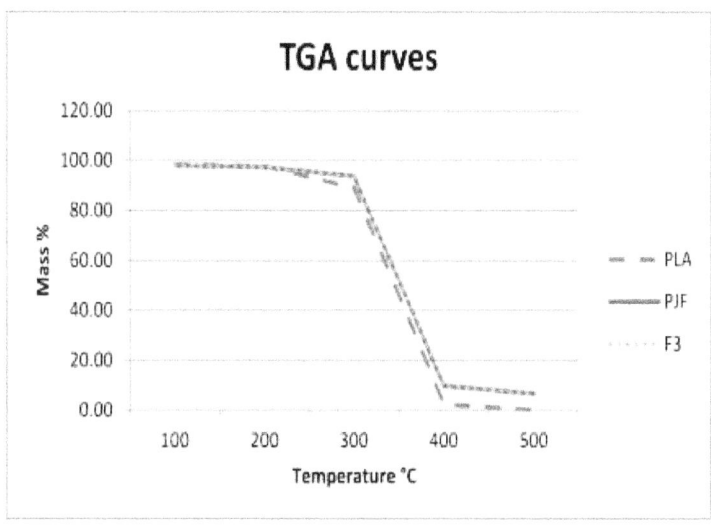

Figure 12.   TGA curves for PLA, PJF and F3.

4. Conclusions

The addition of novel PJF in the form of both coarse and fine fiber reinforcements into PLA had significantly improved the tensile, flexural and impact properties of the polymer. The excellent bonding ability and even distribution of PJF with PLA matrix is the reason behind the improvization of the mechanical properties of the polymer. The PJF PLA composites in both the C and F variants had reduced hardness when compared to the plain PLA sample. The most optimum ratio of reinforcement of PLA to PJF irrespective of the size of particulate that was reinforced was concluded to be 80:20 wt%

On comparing the coarse and fine fiber reinforcements, the fine fiber reinforced composites revealed better mechanical strengths in all the cases of this investigation.

The large lignin content in PJF had greatly supported in improving the stiffness and thermal resistance of the composite material which was proved by the high flexural strength and increased thermal stability respectively. PJF/PLA composite can be sucessfully recomended for the primarily usage and as an alternative for conventional plywood in the construction field.

References

1. Nampoothiri, K. M., Nair, N. R., & John, R. P. (2010). An overview of the recent developments in polylactide (PLA) research. *Bioresource Technology*, vol. 101, 8493-8501.

2. Sykacek, E., Hrabalova, M., Frech, H., & Mundigler, N. (2009). Extrusion of five biopolymers reinforced with increasing wood flour concentration on a production machine, injection moulding and mechanical performance. *Composites: Part A*, vol. 40, 1272-1282.

3. Mofokeng, J. P., Luyt, A. S., Tabi, T., & Kovacs, J. (2011). Comparison of injection moulded, natural fiber composites with PP and PLA as matrices. *Journal of thermoplastic composite materials*, 25(8), 927-948, DOI:10.1177/0892705711423291.

4. Siakeng, R., Jawaid, M., Ariffin, H., Sapuan, S. M., Asim, M., & Saba, N. (2018). Natural fiber reinforced Polylactic acid composites: A Review. *Polymer Composites*, vol. 40, no. 2, 1-18.

5. Garlotta, D. (2001). A literature review of Poly(Lactic Acid). *Journal of Polymers and the Environment*, vol. 9, no. 2, 63-84.

6. Manimaran, P., Senthamaraikannan, P., Sanjay, M. R., & Barile, C. (2017). Comparison of fibres properties of Azarirachta Indica and Acacia Arabica plant for light weight composite applications. *Structural Integrity and Life*, vol. 18, no. 1, 37-43.

7. Saravanakumar, S. S., Kumaravel, A., Nagarajan, T., Sudhakar, P., & Baskaran, R. (2013). Characterization of a novel natural cellulosic fiber from Prosopis Juliflora bark. *Carbohydrate Polymers*, vol. 92, 1928-1933.

8. Tribot, A., Amer, G., Abdou, A. M., Baynast, D. H., Delattre, C., Pons, A., Mathias, J. D., Callois, J. M., Vial, C., Michaud, P., & Dussap, C. G. (2019) Wood lignin: Supply, extraction processes and use as bio-based material *European Polymer Journal*, vol. 112, 228-240.

9. Matuana, L. M., & Stark, N. M. (2015). *The use of wood fibers as reinforcements in composites*. In Omar, F., & Mohini, S. (Eds). Biofiber Reinforcement in Composite Materials. (pp. 648 - 688). DOI:10.1533/9781782421276.5.648, Woodhead publishing United Kingdom, Elsevier.

10. Shimpi, N. G. (2018). *Biodegradable and Biocomposite materials,* Woodhead publishing United Kingdom Elsevier.

11. Petchwattana, N., & Covavisaruch, S. (2014). Mechanical and morphological properties of wood plastic biocomposites prepared from toughened Poly(lacticacid) and rubber wood sawdust (Hevea brasiliensis). *Journal of Bionic Engineering*, vol. 11, 630-637.

12. Wan, L., & Zhang, Y. (2018). Jointly modified mechanical properties and accelerated hydrolytic degradation of PLA by interface reinforcement of PLA-WF. *Journal of the Mechanical Behavior of Biomedical Materials.* vol. 88, 223-230.

13. Guo, R., Ren, Z., Bi, H., Song, Y., & Xu, M. (2018). Effect of toughening agents on the properties of poplar wood flour/poly (lactic acid) composites fabricated with Fused Deposition Modeling. *European Polymer Journal.* vol. 107, 34-45.

14. Huda, M. S., Drzal, L.T., Misra, M., & Mohanty, A. K. (2006). Wood-fiber-reinforced Poly(lactic acid) Composites: Evaluation of the physicomechanical and morphological properties. *Journal of Applied Polymer Science.* vol. 102, 4856-4869.

15. Lee, S. H., & Wang, S. (2006). Biodegradable polymers/bamboo fiber bio composite with bio-based coupling agent. *Composites: Part A* vol. 37, 80-91.

16. Pilla, S., Gong, S., O'Neill, E., Rowell, R., M., & Krzysik, A., M. (2008). Polylactide-Pine wood flour composites. *Polymer Engineering and Science.* vol. 48, no. 3, 578-587.

17. Goud, E. Y., Nagaphani Sastry, M., Devi, K. D., & Raghavendra Roa, H. (2016). Mechanical properties of natural composite fiber Prosopis Juliflora. *International Journal of Innovative Research in Science, Engineering and Technology.* vol. 5, no. 9, 17037-17043.

18. Moyeenudin, A. S., Pickering, K. L., & Fernyhough, A. (2011). Improvement of Mechanical performance of industrial hemp fiber reinforced polylactide biocomposite, *Composites: Part A, 42,* 310–319. DOI:10.1016/j.compositesa.2010.12.004.

19. Orue, A., Eceiza, A., & Arbelaiz, A. (2018). Preperation and characterization of poly lactic acid plasticized with vegetable oils and reinforced with sisal fiber, *Industrial crops and products, 112,* 170-180. DOI: 10.1016/j.indcrop.2017.11.011.

20. Tisserat, B., Joshee, N., Mahapatra, A. K., Selling, G. W., & Finkenstadt, V. L. (2013). Physical and mechanical properties of extruded poly(lactic acid)-based Paulownia elongate biocomposites. *Industrial Crops and Products.* vol. 44, 88-96.

21. Fan, C., Yang, D., Wang, H., Sun, Y., Lou, H., & Yang, H. (2016). Research on preperation methods of ultrafine softwood powder. *International Journal of u⁻ and e⁻ service, Science and Technology.* Vol. 9, no. 4, 225-234.

22. Mazzanti, V., & Mollica, F. (2017). Rheology of wood flour filled Poly(lactic acid). *Proceedings of the Third International Conference on Natural Fibers: Advanced Materials for a Greener World.* 61-67.

23. Chandramohan, D., & John Presin Kumar, A. (2017). Experimental data on the properties of natural fiber particle reinforced polymer composite material. *Data in Brief.* vol 13, 460-468.

24. Sachin, S. R., Kannan, T. K., & Rajasekar, S. (2020). Effect of wood particulate size on the mechanical properties of PLA biocomposite.

Pigment & Resin Technology (ahead of print). DOI 10.1108/PRT-12-2019-0117.

25. Yaacob, N. D., Ismail, H., & Ting, S. S. (2016). Soil burial of Polylactic acid/Paddy straw powder Biocomposite. *BioResources*, vol. 11, no. 1, 1255-1269.

26. Nasrin, R., Biswas, S., Rashid, T. U., Afrin, S., Jahan, R. A., Haque, P., & Rahman, M. M. (2017). Preperation of Chitin-PLA laminated composite for implantable application. *Bioactive Materials*. vol. 2, no. 4, 199-207.

27. Tewari, M., Singh, V. K., Gope, P. C., & Chaudhary, A. K. (2012). Evaluation of mechanical properties of bagasse-glass fiber reinforced composite. *Journal of Materials and Envionmental Science*. vol. 3, no. 1, 171-184.

28. Devaprakasam, D., Hatton, P. V., Mobus, G., & Inkson, B. J. (2008). Effect of microstructure of nano- and micro-particle filled polymer composites on their tribo-mechanical performance. Electron Microscopy and Analysis Group Conference 2007 (EMAG 2007). *Journal of Physics*. DOI: 10.1088/1742-6596/126/1/012057.

29. Nagalingam, R., Sundaram, S., & Retnam, B. S. J. (2010). Effect of nanoparticles on tensile, impact and fatigue properties of fibre reinforced plastics. *Bulletin of Materials Science*. vol. 33, no. 5, 525-528.

30. Lan, H., & Venkatesh, T. A. (2014). On the relationships between hardness and the elastic and plastic properties of isentropic power-law hardening materials. *Philosophical Magazine*. vol.94, no. 1, 35-55.

31. Gacitua, W., Bahr, D., & Wolcott, M. (2010). Damage of the cell wall during extrusion and injection molding of wood plastic composites. *Composites: Part A*, vol. 41, pp. 1454-1460.

32. Nampoothiri, K. M., Nair, N. R., & John, R. P. (2010). An overview of the recent developments in polylactide (PLA) research. *Bioresource Technology.* vol. 101, pp. 8493-8501.

33. Raj, S. S., Kannan, T. K., Babu, M., & Vairavel, M. (2019). Processing and testing parameters of PLA reinforced with natural plant fiber composite materials – A brief review. *International Journal of Mechanical and Production Engineering Research and Development*, vol. 9, no. 2, 933-940.

34. Sifat, R., Akter, M., & Rashid, A. K. M. B. (2016). Properties of micro-nano particle size admixtures of alumina at different sintering condition. *International Conference on Mechanical Engineering*, AIP conference proceedings 1754, 030005 1-6, DOI: 10.1063/1.4958349.

35. Chaparro, T. D. C. (2016). *Synthesis of nanocomposites with anisotropic properties bby controlled radical emulsion polymerization lorena.* (Thesis in Material Chemistry). University of Lyon, France.

36. Balart, J. F., Fombuena, V., Fenollar, O., Boronat, T., & Sanchez Nacher, L. (2016). Processing and characterization of high environmental efficiency composites based on PLA and hazelnut shell flour (HSF) with biobased plasticizers derived from epoxidized linseed oil (ELO). *Composites Part B: Engineering*, vol. 86, 168-177.

37. Binoj, J. S., Edwin Raj, R., & Daniel, B. S. S. (2017). Comprehensive characterization of industrially discarded fruit fiber, Tamarindus Indica L. as a potential eco-friendly bio-reinforcement for polymer composite. *Journal of Cleaner Production.* vol. 142, no. 3, pp. 1321-1331.

Publisher: Eliva Press SRL

Email: info@elivapress.com

Eliva Press is an independent publishing house established for the publication and dissemination of academic works all over the world. Company provides high quality and professional service for all of our authors.

Our Services:
Free of charge, open-minded, eco-friendly, innovational.

-Free standard publishing services (manuscript review, step-by-step book preparation, publication, distribution, and marketing).
-No financial risk. The author is not obliged to pay any hidden fees for publication.
-Editors. Dedicated editors will assist step by step through the projects.
-Money paid to the author for every book sold. Up to 50% royalties guaranteed.
-ISBN (International Standard Book Number). We assign a unique ISBN to every Eliva Press book.
-Digital archive storage. Books will be available online for a long time. We don't need to have a stock of our titles. No unsold copies. Eliva Press uses environment friendly print on demand technology that limits the needs of publishing business. We care about environment and share these principles with our customers.
-Cover design. Cover art is designed by a professional designer.
-Worldwide distribution. We continue expanding our distribution channels to make sure that all readers have access to our books.

www.elivapress.com

www.ingramcontent.com/pod-product-compliance
Lightning Source LLC
Chambersburg PA
CBHW051303170526
45165CB00004B/1836